CLIFF RAILWAYS, LIFTS AND FUNICULARS

Martin Easdown

AMBERLEY

A Lynton and Lynmouth Lift car ascends from the lower station in Lynmouth, August 2017.

For Linda, who has loyally travelled with me
through life's ups and downs.

First published 2018

Amberley Publishing
The Hill, Stroud
Gloucestershire, GL5 4EP

www.amberley-books.com

British Library Cataloguing in Publication Data.
A catalogue record for this book is available from the British Library.

ISBN 978 1 4456 8003 3 (print)
ISBN 978 1 4456 8004 0 (ebook)

Origination by Amberley Publishing.
Printed in Great Britain.

Introduction

Often associated with the British seaside, the cliff lift has been a much-loved, and rather eccentric, method of transporting people up and down a cliff side for nearly 150 years. They are otherwise known as cliff railways, incline tramways or funicular railways (the terms have always been interchangeable, as they are in this publication). The word 'funicular' is defined as being of 'rope or tension': in other words, a cable-hauled railway or tramway worked by a system of hydraulic balance and running on their own private ground. Cliff railways are directly descended from cable-hauled railways, prevalent in mines and quarries, and also from early passenger railways, where an engine or winding gear hauled loads up steep slopes. The first European funicular railway was built at Lyon, France, in 1862 and this was followed by the Budapest Castle Hill Railway in 1868. The first British cliff railway was the South Cliff Tramway at Scarborough in 1875 – the first of five to be built in the resort.

The working operation of a funicular railway was simple, but ingenious. Originally, many of the passenger cars boasted a water tank beneath the undercarriage that was filled at the upper station with water from a piped supply. Once the brake was let off, the water-filled top car descended the cliff, and by means of a strong connecting steel cable, its weight pulled up the lighter bottom car. At the end of the journey, the water was drained and emptied into a sump tank before being pumped back up to a tank at the top and then into the undercarriage tank of the car before the cycle was ready to repeat itself once more. The pumps were originally powered by gas or oil engines, which were later replaced by electric motors. The water supply and application of the brakes were controlled by a brakeman, who could stop the cars in an emergency, such as the slackening of tension in the main cable or the unlikely event of it breaking. This was catered for by fitting an extra small cable, which would bring to bear reversed steel quadrants to brake the cars. More modern safety features included trackside switches and over-speed governors. Many of the Victorian water-balance lifts were later converted to electric operation, which was the preferred method of traction of those lifts built during the twentieth century. However, three of the nineteenth-century water-balance lifts, at Saltburn-by-the-Sea, Folkestone and Lynton and Lynmouth, survive, and they were joined in 1992 by a new water-balance lift at the Centre for Alternative Technology near Machynlleth.

As is the case with railways (Brunel and Stephenson) and piers (Birch and Brunlees), cliff railways have their own iconic names associated with them. The partnership of financier Sir George Newnes (1851–1910) and George Croydon Marks (1858–1938) were responsible together or separately for several cliff railways, including Lynton and Lynmouth (1890), Bridgnorth (1892), Clifton Rocks (1893) and Aberystwyth (1896). They also built the incline cable tramways at Matlock (1893) and Swansea (1898), and Newnes paid for the vertical lift at Shanklin (1893). Newnes was a successful publisher (his publications include *Titbits* magazine and the Sherlock Holmes stories) and the Member of Parliament for Newmarket in 1891–95 before being created a Baronet in 1895. Croydon Marks worked for several renowned engineering firms before joining Tangye Limited of Birmingham, where he became an acknowledged expert in hydraulic machinery and supplied the equipment for the Saltburn Cliff Railway in 1884. Three years later he set up as a consulting engineer in Birmingham, and in 1889 went into partnership with the inventor of the two-stroke engine, Dugald Clerk. As well as being involved with other British cliff lift concerns, Croydon Marks was appointed consulting engineer for a funicular erected in Budapest in 1894. He patented several safety devices for his lifts, including a hydraulic braking system, an automatic brake using steel wedges on the running track and cars with a brakeman on board using a dead man's handle operation, as on the railways.

Another name associated with cliff lifts was R. Waygood & Company, the supplier of hydraulic and electric lifts. The company was founded by Richard Waygood in the 1830s and by the late Victorian era they were the principal installers of lifts in the United Kingdom and had installed them in Windsor Castle and Buckingham Palace, as well as in many other houses, shops and commercial premises. In 1900, Waygood became a public liability company, and two years later a tentative merger with Otis Elevators was unsuccessful, although Waygood did become part of the Otis group in 1914.

This book features all the cliff lifts/railways that have transported people up and down a cliff and mountainside in Britain that are or were regularly open to the public. Not included are some of the recently installed funiculars in private tourist attractions such as Legoland and the National Coal Mining Museum, or the Skyrail at Birmingham Airport, which is a horizontal funicular! Seaside elevator lifts (often termed 'cliff lifts') are included, as are cable tramways that were associated with their cliff railway cousins, such as the ones provided by Newnes and Croydon Marks.

Cable tramways were powered by a continuously moving steel cable buried in a conduit placed centrally between the track. An open slot in the conduit allowed a gripper hanging from the tramcar access to the cable conduit to propel the car. Releasing the gripper from the cable disconnected the power and brought the tram to a stop. The pioneer cable tramway was installed in San Francisco in 1873 and the success of the system led to other cable tramways being installed in American cities, and pioneer (and relatively short-lived) British systems in Birmingham and London, along with the longer-lasting Edinburgh system. By the 1890s, the American cable tramways were being superseded by those with electrical traction, but in Britain two new cable tramways, at Matlock and Douglas, were opened, although they were both closed in the 1920s. In the early years of the twentieth century the Llandudno Great Orme Tramway was opened, which is the only surviving cable tramway in the United Kingdom.

I hope you will enjoy this written journey around Britain's cliff lifts, railways and funiculars – a form of transport with a unique charm of its own.

Martin Easdown
January 2018

Aberystwyth Cliff Railway 1896

Standing high above the north end of Aberystwyth's curved seafront promenade, Constitution Hill in the late Victorian era was a popular pleasure ground known as 'Luna Park'. The attractions included a switchback railway, camera obscura and an open-air dance platform. In May 1895 the Aberystwyth Improvement Company was formed to take over a project by electrical engineers Messrs Bourne & Grant to improve access to the top of the hill by providing a cliff railway. The team of financier George Newnes and engineer George Croydon Marks oversaw the project and the construction work commenced in October 1895. The gangs of workers had a hard task, removing 12,000 tons of rock and shale from the hillside and creating a cutting for the railway to carry existing footpaths over it. Additional attractions were also provided on the summit of the hill in the form of a pavilion, summer houses, a bandstand and flower gardens. The total cost of the work was £60,000 and on Saturday 1 August 1896 the cliff railway was opened to the public to the accompaniment of the Newton Silver Band. 520 people paid the rather pricy sum of 8d to travel on the first day (the fare was subsequently reduced to 3d for an ascent and 2d for a descent). The day's proceedings were concluded with an evening confetti battle in the illuminated gardens.

The double track railway had a gauge of 4 feet 10 inches and undulated up and down the hill for 778 feet (238 metres) on a gradient of 1 in 2 using the water-balance system. Water tanks were installed under the floor of the bottom station to receive the water drained from the descending car. A coal-fired boiler provided the power for the Worthington compound steam pumps to pump the water back up to the top tank under the bandstand. The two cars had stepped seating and were constructed by R. Jones & Sons of Lynton.

In 1921, the railway was converted to electric working when a Morley 55 hp motor was installed at the upper station to wind the cars. The 450 V DC current was taken directly from the town's power station. By this time, most of the attractions at the top of Constitution Hill had been cleared away, although the railway remained busy, carrying those who enjoyed a stroll on the hill to enjoy the spectacular views obtained from it.

The cliff railway changed hands in 1948 when it passed to the Aberystwyth Cliff Railway Company, a subsidiary of the Aberystwyth Pier Company. Further changes of ownership occurred in 1970 and 1976 before the railway was acquired by Constitution Hill Limited, a community interest company, in 1998.

The re-instatement of attractions on Constitution Hill commenced in 1985 when a new camera obscura, the world's largest with a 14-inch diameter lens, was installed. In 2005 a café/restaurant and gift shop were built thanks to Objective One funding. Ten years later, the top floor of the impressive lower station was converted into a holiday apartment.

The railway is normally open from April to October and has a journey time of two minutes and twenty seconds, with the cars travelling at 4 miles per hour. The two cars can seat a total of thirty passengers and are one of only two pairs of stepped cars still operating. They are named *Lord Marks* and *Lord Geraint*.

Above left: A postcard view of the Aberystwyth Cliff Railway in the 1930s, looking up towards the upper station.

Above right: Looking down towards the town of Aberystwyth from Constitution Hill in the 1960s, with the cliff railway in the foreground.

Right: The Aberystwyth Cliff Railway and its impressive lower station, photographed in the 1960s.

Below: The Aberystwyth Cliff Railway car *Lord Marks*, pictured at the lower station in September 2005.

Blackpool Cabin Lift 1930

One of Blackpool's lesser-known seaside attractions, the Cabin Lift on the North Shore is an attractive neo-classical building with a distinctive pyramidal roof. Designed by Borough Architect John Charles Robinson and opened in 1930, this elevator lift was provided to give easier access to the promenade and boating pool. The pool was opened in 1923 and was run by the Maxwell family, who introduced some of the first paddle boats into the country.

In 1979 the lift was closed due to the cliff subsidence, which caused the lift shaft to twist and the bridge link to settle. For the next ten years the lift remained closed while the council decided what to do with it. Eventually, in October 1989, work commenced to restore the lift and it was reopened in March 1991 at a cost of £85,000. The attractive classical styling was refurbished, although the waiting room on the bridge link at the upper entrance was not rebuilt. A lift attendant was employed for two seasons before being deemed to not be financially viable. In March 2010, the lift was given Grade II listed status, but is currently not in use. The former boating pool was filled in and in recent times had a go-kart track laid on it.

Above left: The Boating Pool and Cliff Lift, Blackpool, featured on a postcard from the 1960s.

Above right: The ornate features of the Blackpool Cabin Lift, photographed in September 2001.

Bournemouth East Cliff Lift 1908

The Borough of Bournemouth has three cliff lifts, all of which are owned by the council. The oldest and, until its recent enforced closure, busiest, was the East Cliff Lift, opened on 16 April 1908 by Lady Meyrick. The lift was built by R. Waygood (track and machinery) and Harrison & Company (earthworks and toll house) at a cost of £3,500 and was electrically driven by a 25 hp motor using mains electricity converted to 500 V DC. The motor and winding gear were placed in the upper station, located on East Overcliff Drive. The track, of 5 foot 6 inch gauge, runs for 170 feet (51 metres) on a gradient of 1 in 1.46.

All three Bournemouth lifts were upgraded on a regular basis. The East Cliff Lift received new cars in 1960, seating twelve passengers, which were designed to be interchangeable with the chassis on all three lifts. In 2007 new stainless steel cars were introduced. The track received an overhaul in 1987 when new rails and pre-cast concrete units replaced the original timber baulks, and during the 1990s the lifts received an electronic control system.

However, on 24 April 2016, a landslip of the cliff significantly damaged the track and lower station, resulting in the closure of the lift for the foreseeable future.

A commemorative postcard of the opening of the Bournemouth East Cliff Lift on 16 April 1908.

East Cliff Lift, Bournemouth.

Left: A postcard by W. H. Smith & Son of the East Cliff Lift, Bournemouth, *c.* 1910.

Below: Looking down the incline of the East Cliff Lift, Bournemouth, in October 2013.

Bournemouth West Cliff Lift 1908

Bournemouth's second lift, on the West Cliff, was opened by the corporation on 1 August 1908 and resembled its counterpart on the East Cliff. However, it was shorter, at 145 feet (44 metres), and had a rise of 102 feet on a gradient of 1 in 1.42. The track gauge, 5 feet 6 inches, was the same as the East Cliff Lift and so were the contractors: Waygood (track and machinery) and Harrison (earthworks and lower station). The lift was electrically powered from the outset. The total cost of construction was £4,395.

Although lesser used than its East Cliff counterpart, the West Cliff Lift gave sterling service and in the 1960s was overhauled, when aluminium cars were installed, along with a new 28 hp three-phase winding motor. The cars were replaced again in 1975 (with the capacity reduced from sixteen to twelve persons in each car), and in 1987 the track was re-laid. During the 1990s an electronic control system was installed, operated from the upper station.

The lift is normally open from April to October and has a journey time of forty seconds.

The West Cliff Lift, Bournemouth, photographed by Harvey Barton *c.* 1910.

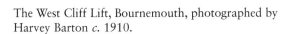

The Lift, West Cliff, Bournemouth

A Valentines postcard of the Bournemouth West Cliff Lift in operation soon after it was opened in 1908, with the bare cliff side clearly noticeable.

A photograph of the Bournemouth West Cliff Lift in October 2013, showing the lower station looking much the same as when first built, although it has been extended.

Bournemouth Fisherman's Walk Lift 1935

Bournemouth's third lift, at Fisherman's Walk, is situated in the suburb of Southbourne (once a separate resort of its own) and is handily placed for both the Southbourne and Boscombe promenades. The lift was designed by the Borough Engineer, F. P. Delamore, and was opened on 8 June 1935 at a cost of £6,925. It runs for a length of 128 feet (39 metres), with a rise of 86 feet on a gradient of 1 in 1.49. The cars were supplied by the Express Lift Company, as was the 21 hp 500 V DC electric winding motor. This was replaced by a three-phase 415 V motor in the 1960s. New cars were introduced in 1978, which have a passenger capacity of twelve. They run on a track gauge of 5 feet 8 inches – 2 inches wider than the other two Bournemouth lifts.

The lift is normally open April to October and it has a journey time of fifty seconds.

Above left: The seafront at Southbourne in the 1950s, showing the Fisherman's Walk Lift, which was opened in 1935.

Above right: The Fisherman's Walk Lift, photographed on a sunny spring day in March 2004.

Bridgnorth Cliff Railway 1892

Also known as the Castle Hill Railway, the inland cliff railway at Bridgnorth is 201 feet (61 metres) long and connects High Town with Low Town. The railway has a rise of 111 feet on a gradient of 1 in 1.8, and the 3 foot 8½ inch gauge double track was laid in a 50-foot-deep cutting in the sandstone rock cliff.

The proposals for a cliff railway were aired at a public meeting in 1890 to discuss means of connecting High and Low towns to avoid a climb of 200 steps, and in the following year the Bridgnorth Castle Hill Railway was registered with a capital of £6,000 in £10 shares. The two leading lights in the company were the lift builders supreme George Newnes and George Croydon Marks. The former provided most of the finance, while the latter was the line's engineer and its first managing director (until 1901). Marks' brother, Edward, was also involved in the company, and became its managing director between 1901 and 1924. George Law of Kidderminster was appointed the contractor and commenced work on 2 November 1891. Good progress was made, and the cliff railway was opened on 7 July 1892.

The line was worked on the water-balance principle and the original cars had two longitudinal seats each seating nine people. These sat on a horizontal platform at the top of a triangular steel frame, which carried a 2,000-gallon water tank. The cars were fitted with automatic gripper brakes, which were to be used if they were travelling too fast. In addition, an automatic wedge brake, held off by the tension of the cables, would activate if the cable broke. As a further safety measure, a brakeman travelled on each car on a small platform. Two Otto gas engines pumped the water back up to the upper station from the basement reservoir, where the cars could be filled from a 30,000-gallon tank on the roof of the red brick station building.

The railway proved to be a useful asset, but by the 1920s the Bridgnorth Castle Hill Railway were struggling financially. In April 1933 they closed the line and it lay dormant until May 1934, when it was reopened by a reconstituted Bridgnorth Castle Hill Railway led by Frank Myatt. The Myatt family were to remain owners of the company until 1996.

Sadly, a fatality occurred on the railway in May 1943 during work to change the steel wire cables when the upper car ran down the hill, crushing the engineering manager, Mr Howes, against the lower station. During the reconstruction work, the decision was taken to convert the railway to electrical haulage, which was carried out by

Messrs Francis and J. S. Lane, with the electrical gear provided by the Metropolitan-Vickers Electrical Company of Manchester. The winding gear was installed at the upper station and was controlled by a single operator, who also sold the tickets. The brakeman who rode on the cars was dispensed with. In 1955, the original car bodies were replaced by streamlined aluminium bodies resembling a motor coach. These were provided by F. S. Fildes of Stourbridge at a cost of £1,790 and could each seat eighteen people.

The Bridgnorth Castle Hill Railway still operate the railway, and since 2011 it has been owned by relatives of George Croydon Marks, who keep the line running 362 days of the year.

Right: An early postcard view showing the Bridgnorth Castle Hill Railway *c.* 1907. Note the brakemen standing on the outside platform of the cars.

Below: The platforms of the cars had been enclosed by the time of this postcard, which shows the upper station on Castle Walk in the 1920s.

Above: A view of the upper station in the 1990s, showing the streamlined coach-like cars introduced in 1955.

Left: Bridgnorth as seen from the top of the cliff railway in 2007.

Brighton Madeira Drive Lift 1890

This is an attractive electrically operated lift that has an oriental design in keeping with Brighton's most famous building, the Royal Pavilion. The lift was built as part of the Madeira Terrace development of a covered promenade along Madeira Road (now Madeira Drive), which was opened on Saturday 24 May 1890 after the completion of the work by Messrs Longley & Sons of Crawley at a cost of £28,000.

Borough Engineer Philip Lockwood's design for the upper station on Marine Parade featured an elegantly constructed kiosk of Indian teak with an ornamental domed roof of copper. A griffin decorated each of the four corners, along with four dolphins, and an iron weathervane surmounted the top. The lower access to the lift was reached through a shelter hall used as a refreshment and concert room. The Hydraulic Engineering Company of Chester provided the lift equipment, which was originally powered by water from the town's mains. Up to 90 gallons was used each time the lift cage was raised. The fare was a ha'penny and up to fifteen people could be carried.

The lift has suffered periods of closure, and in 2013 it was reopened after a £200,000 restoration, which included renovating the roof, installing night lighting and refurbishing the lift mechanism. It is a Grade II listed building and privately operated between April and October on behalf of Brighton & Hove City Council. The lift is now limited to carrying six people at a time and is attendant operated.

An attractive postcard showing the lift and its associated shelter hall and Madeira Road covered promenade c. 1905.

The Lift and Madeira Road, Brighton

The upper station of the Brighton Madeira Drive Lift on Marine Parade, photographed in August 2017.

The shelter hall on Madeira Drive, which doubles as the lower station of the lift, photographed in August 2017.

Bristol Clifton Rocks Railway 1893–1934

The Clifton Rocks Railway was another George Newnes and George Croydon Marks instigated cliff railway and had four tracks that ran through a 500-foot tunnel in the cliffs of the Avon Gorge. The Society of Merchant Ventures, owners of the Avon Gorge cliffs, had received a proposal from Newnes on 26 September 1890 to build an incline railway from Hotwells (not far from the Bristol end of Brunel's suspension bridge) through a tunnel in the cliffs to emerge in the garden of 14 Princes Buildings, Clifton. The proposal was agreed, provided Newnes provided a hydropathic spa next to the upper station of the railway. Architects appointed for the project were Philip Monroe & Sons of Baldwin Street, Bristol, and the contractors were Messrs C. A. Hayes of Bristol. Work was commenced in 1891, but problems with faults in the limestone rock strata meant that the tunnel had to be brick-lined, tripling the estimated cost of £10,000. With a roof height of 18 feet and a width of 27 feet 6 inches, the tunnel was at the time the widest in the world, and climbed a vertical distance of 240 feet on a rising gradient of 1 in 1.2. The track gauge was 3 feet 8 inches and the four cars were built by Messrs Gimsons of Leicester and could each seat eighteen passengers. There was also an outside platform for the attendant. The cars were worked on the water-balance method, with the water emptied into the storage tanks at the bottom returned to the top by a pair of pumps powered by four-stroke Otto gas engines. The upper station was situated in a triangle of land at the junction of Princes Lane and Sion Hill, from where passengers descended a flight of steps to reach the platform. A cab in the station housed the brakeman. The lower station, housing the pumping machinery, was constructed inside the rock with its façade erected flush with the rock face.

The railway was finally opened on 11 March 1893, when 6,200 people made the return journey, each receiving a gilded medallion. The opening day fare of 4d was then reduced to 1d up and ½d down, which was increased in 1906 to 2d up and 1d down.

Having personally financed the cost of the railway, Newnes formed the Clifton Rocks Railway Company in the spring of 1894, with himself as chairman and Croydon Marks as a director. The same year saw Newnes deliver the promised spa, which adjoined the upper station of the railway. Named the Clifton Spa Pump Room, and designed by Philip Monroe, water was drawn via a 350-foot borehole from Hotwells below, where a spa had flourished during the seventeenth and eighteenth centuries. The Clifton Grand Spa Hydro (now the Avon Gorge Hotel) was opened next to the pump room in 1898.

The Clifton Rocks Railway got off to a good start, with 427,492 passengers being carried in the first year of operation, but visitor numbers steadily declined thereafter and in 1908 the Clifton Rocks Railway went into receivership. In 1912, the railway was purchased by the Bristol Tramways & Carriage Company for £1,500, yet passenger numbers declined even further following the closure of Hotwells station on the Bristol Port & Pier Railway in 1922 and the company decided to close the Clifton Rocks Railway on 29 September 1934.

But that was not the end of the story. During the Second World War, the tunnel came back to life when it was used by the BBC to house its wartime emergency headquarters, and by British Overseas Airways for office accommodation and storage. The BBC withdrew from the lower section of the tunnel in 1955 (but remained in the upper section until 1960), and in the following year, a 4-inch-wide crack was discovered between the lower station façade and the limestone cliffs. Buttresses had to be constructed in 1958 to stabilise the façade.

The Clifton Rocks Railway Group occasionally opens the tunnel to the public, where there is still much evidence of the BBC's occupation and some remains of the railway. The group have carried out restoration work and the long-term aim is to restore and operate the Clifton Rocks Railway as a sustainable visitor attraction.

Above left: The lower station of the Clifton Rocks Railway is the subject of this postcard sent in 1904. Note the ice cream seller, and the proximity of Brunel's suspension bridge.

Above right: An interior view of the Clifton Rocks Railway *c.* 1904, showing the four tracks and open platform of the car, a feature of the Newnes-Croydon Marks cliff lifts.

The upper station of the Clifton Rocks Railway as restored by the Clifton Rocks Railway Group, photographed in August 2017. The Avon Gorge Hotel can be seen on the left.

The disfigured lower station entrance on Hotwells Road, photographed in August 2017.

Broadstairs Viking Bay Lift 1 1910–91

An interesting, but rarely photographed lift, which ran from the back garden of 14 Albion Street, Broadstairs through the chalk face of the cliff to emerge on the Viking Bay beach. The line ran for 100 feet (30 metres) on a gradient of 1 in 1.41 and had a track gauge of 5 feet 3 inches. The single car, which could carry up to twelve passengers, was electrically driven and counterbalanced with a large concrete weight.

The lift, which travelled through a brick-lined shaft, was built by G. Graham Tucker of Ramsgate. Waygood provided the lift mechanism and operated the lift through its subsidiary Cliff Lifts Limited. In 1920, Mrs S. E. Wilson of Cambridge Villa, Broadstairs took over the running of the lift, and upon her death it passed to Miss M. A. Wilson. The next owner was the Garrett family, who ran the venture to the outbreak of the Second World War, when it was closed. By the end of the war, the lift needed extensive work done and it was sold to Thanet Amusements Limited, who repaired and reopened it. However, the lift changed hands again in 1951 and 1957. By 1960, the fare was 4d for adults and 2d for children under twelve years of age and season tickets were 5s weekly and 25s for the season. In 1963 Mr F. J. Langrete of Seaview Road, Broadstairs, became the next owner and gamely kept the lift running, but by 1977 it was in a poor state of repair and closure appeared imminent. The Broadstairs Lift Company was formed to run the lift, although it was largely down to Ken Kneeshaw that it managed to remain open until 1991, when an electrical fault finally brought about closure.

The lower station can still be seen fronting the beach, but the upper station was removed to make way for a patio for a restaurant, although this does incorporate an access door to the lift shaft.

A postcard of a busy Viking Bay, Broadstairs, in the 1930s. The lower entrance to the lift, with its white roof and lettering above, can just been seen centre right of the picture.

Right: The lower entrance to the Viking Bay Lift, Broadstairs, photographed in the early 1960s. The attendant can be seen looking at the photographer.

Below: The upper station of the lift was built in the back garden of 14 Albion Street. It can be seen here in the early 1960s, with plenty of signs pointing the way to it!

Broadstairs Viking Bay Lift 2 2000

Situated just along from the closed cliff railway at Viking Bay, this vertical elevator lift was opened by Thanet District Council in 2000. The attractive structure is 53 feet (16.2 metres) high, brick-built and rendered and painted. It features picked-out brickwork around the entrance doors and beside the full-height window with flint work at high level. The words 'LIFT' are picked out on both sides of the shaft, and '2000' can be viewed on the seaward side. A bridge provides access to the upper entrance on Victoria Gardens, while the lower entrance leads straight on to the sands. The lift motor room is in a storeroom within the adjoining chalet block. The lift car has a capacity of fourteen persons and is free to use.

Two views taken in September 2017 of the lift built in 2000 at Broadstairs.

Cairngorm Mountain Railway 2001

The Cairngorm Mountain Railway is Scotland's only funicular railway and Britain's highest and longest. The railway is located within the Cairngorms National Park, the largest national nature reserve in Britain, which is eight miles from Aviemore. It is owned by Highlands & Islands Enterprise and was operated by Cairngorm Mountain Railway Limited until July 2014 when Natural Retreats acquired the lease.

Construction of the railway began in August 1999 to a design by civil engineers A. F. Cruden Associates of Inverness and it was opened on 24 December 2001. The railway replaced a chairlift, which was too sensitive to strong winds, and runs for 6,460 feet (1,969 metres) up the mountain to the upper station, where, aside from the spectacular panoramic views, there is a visitor centre, restaurant and shop. The single track has a gauge of 6 feet 6½ inches and runs on a maximum gradient of 1 in 2.5. There is a passing loop just above the middle station. Up to sixty passengers can be carried in each of the two totally enclosed cars, which are accessible to wheelchair users. The system is powered to two 500 KW electric motors and a 'soft-start, soft-stop' control system to regulate the car speeds. A hydraulically operated 'counter' cable is used in addition to the main hauling cable to maintain haul-rope tension.

The line is open throughout the year except Christmas Day, and during the winter the cars travel at a quicker speed to get skiers up the mountain in just five minutes. The more leisurely summer journeys take nine minutes, although passengers can only access the mountain on a guided tour.

The Cairngorm Mountain Railway, and the spectacular scenery through which it runs, in 2016.

Clacton-on-Sea Reno Electric Stairway 1902–8

Developed by the American Jessie Reno in 1892, his moving stairways (otherwise known as inclined elevators) had been installed in larger American department stores by the turn of the twentieth century and at the New York playground of Coney Island. Five were commissioned for the 1900 Paris Exposition and one was installed in London's Crystal Palace to transport patrons from the ground floor in the central transept to the gallery for a penny a ride. In 1901, Reno installed a moving stairway at Seaforth Sands station on the Liverpool Overhead Railway (which lasted for about five years), and others at the Earl's Court Exhibition in London, Alhambra in Blackpool and at Southend-on-Sea. He moved to Britain to oversee his ventures and in 1902 the Reno Electric Stairways & Conveyors Company Limited was formed. They opened a stairway in the resort of Clacton-on-Sea, where the electric-powered escalator transported its passengers 40 feet up the cliff for a penny a ride.

The novelty factor of the stairway ensured it was initially successful, but it was soon losing money as it was easier and often quicker to walk up and down the cliff paths. In 1906 the Reno company wanted to remove the ride, and following legal wrangling with the council this was achieved in 1908.

In 1911, Otis Elevators (who had bought a third share in Reno's British company in 1903) acquired Reno's patents and in the following year the Reno Electric Stairways & Conveyors Company was dissolved.

Below left: The Reno Electric Stairway at Clacton-on-Sea can be seen centre right on this postcard from 1907, climbing up the cliff side.

Below right: A family relaxes on the beach at Clacton, and behind them can be seen the Reno Electric Stairway. This postcard was also sent in 1907, one year before the stairway was removed.

Devil's Dyke Steep Grade Railway 1897–1908

The spectacular Devil's Dyke on the South Downs above Brighton, which boasts the longest, deepest and widest dry valley (or gorge) in the UK, was a popular day trip destination during the Victorian period. The Brighton & Dyke Railway was opened on 1 September 1887 and the area around the Dyke Hotel was opened from 1892 as a pleasure ground by James Hubbard. On Whit Monday 1893, Hubbard attracted around 30,000 visitors to the Dyke, and among the amusements added to entertain them were a switchback railway, Hotchkiss Bicycle Railway, swings and a steam-powered roundabout. To capitalize on his success, Hubbard engaged civil engineer William Brewer to design methods of transportation to the hotel. The original idea of a cable tramway from Dyke railway station to the hotel failed to materialise, but in October 1894 an aerial cableway across the Dyke gorge was opened. Built by Cable Tramways Construction and erected by Heenan & Froude at a cost of £5,000, the cableway consisted of two pylons 650 feet apart from which the cable was suspended 230 feet above the Dyke. A small cage work car, powered by a Crossley's patent oil engine and carrying four passengers, operated between the north and south stations. For a single fare of sixpence, the passengers could experience the thrill of a two-and-a-half-minute aerial flight across the Dyke.

Three years later, on 24 July 1897, a funicular railway was added to the Dyke's attractions. Hubbard had leased the land for its construction to the Pyramidical Syndicate Limited of London at a ground rent of £100 per annum, although the line was to be owned and operated by the Brighton Dyke Steep Grade Railway Limited. They had been formed with a capital of £10,000, from which £9,000 had to be raised for the railway's construction, as per the agreement with the Pyramidical Syndicate. Engineered by Charles Blaber, construction of the railway was placed in the hands of Messrs Courtney & Birkett of Southwick, while the Ashbury Railway Carriage & Iron Company were to build the two cars. These were designed to be open above the waist rail and have a platform at each end for the conductor/brakeman. Sliding doors accessed the cars at both ends and the floors were at an angle of 30 degrees to the horizontal. The line was laid down the Shepherd's Steps Combe on the escarpment to the north-east of the hotel and was handy to the visit the village of Poynings and its tea rooms. It was also hoped that the Dyke Hotel could be supplied with produce from local farms. The two 3 foot gauge tracks were laid to a length of 840 feet (256 metres) on a rise of

395 feet and the steepest gradient of 1 in 1.5. The two cars could carry up to fifteen passengers each (twelve seated) at a speed of 3 miles per hour and were hauled by steel cables around a pulley wheel driven by a 25 hp Hornsby-Ackroyd oil engine. This was in the upper station, which consisted of a low brick building with open platforms on trestles. The lower station had no building at all (although a grand one was proposed) and consisted of just two open platforms.

The Steep Grade Railway got off to a good start and it was claimed that around 275,000 passengers per year had been carried. However, during 1899, the railway ceased running due to the financial woes of its operating company. It was revealed that the company owed £4,766 to the Pyramidical Syndicate Limited, who were being wound up, and on 16 July 1900 the railway was put up for auction. It failed to sell, and was offered again at another auction on 13 December 1900, but was withdrawn after the highest bid only reached £350. A syndicate of debenture and shareholders of the Brighton Dyke Steep Grade Railway acquired the railway and then the company was wound up in 1902.

The early years of the new century saw a downturn in fortunes of the Dyke Estate. The switchback and bicycle railways had ceased running and the cliff railway was continuing to lose money. So was James Hubbard, who overstretched his finances and fled to Canada in 1907. The Steep Grade Railway closed in 1908 and the aerial cableway ran for the last time the following year. The equipment of the cliff railway was removed around 1913, but the upper station survived into the 1920s and its foundations can still be seen, as can the concrete bases of the aerial cableway towers.

Above left: A postcard by the Mezzotint Company of Brighton showing the Devil's Dyke Steep Grade Railway in 1904. The total absence of a lower station at the Poynings end can clearly be seen.

Above right: Another Mezzotint postcard, showing the Devil's Dyke end of the Steep Grade Railway with its engine house. One of the cars can be seen to the left of the platforms.

No. 135. Devil's Dyke near Brighton. — The Steep Grade Railway, constructed by the Landlord of the Dyke Hotel to take passengers from the summit of the hill down to the level of Poynings, a pretty little Sussex Village. This view shows the Cleft or Combe which is called the Devil's Dyke.

Another view of the upper station of the Devil's Dyke Steep Grade Railway from about 1904, four years before it was closed.

A postcard from around 1905 showing the aerial cableway, which provided a spectacular ride across the Dyke gorge between 1894 and 1909.

Douglas Falcon Cliff Hotel Lift 1 1887–96

The Falcon Cliff is an imposing building standing high above Douglas Bay that was erected in the 1840s for John Stanway Jackson, the manager of the Bank of Mona. In 1877, the house was converted into a hotel and amusements were added in a bid to rival the successful Derby Castle complex down on the foreshore. These included a skating rink, and in 1887 a large glass dance pavilion was erected. At the same time, it was decided to improve the access to the hotel and its entertainments by providing a cliff lift. T. Cain was employed to carry out the work, but tragedy struck on 30 May 1887 when Cain's son, Alfred, was killed by an accident on the site. Nevertheless, Cain pressed on with the work and on 6 August 1887 the lift was opened. It ran for 218 feet (66 metres), with a rise of 110 feet and a gradient of 1 in 1.98 on a double track of 4 foot gauge. The lift was powered by an oil engine and took one minute and twenty-five seconds to complete its journey.

Although it was a useful addition for the hotel's patrons, and saved a steep walk up to the hotel complex, by the 1890s the glut of large entertainment centres in Douglas was affecting the fortunes of the Falcon Cliff, and in 1896 it was decided to close both the lift and dance pavilion. The latter was demolished but the lift was acquired by the Forrester brothers, who re-erected it at Port Soderick in 1898 to give easier access to their own entertainment complex. That same year saw the sensible merger of the Falcon Cliff with the Derby Castle and Palace concerns.

The first Falcon Cliff Hotel Lift at Douglas, photographed *c.* 1890. Of note is the castle-like entrance building to the lift (matching the hotel above), and the huge glass dance pavilion to the left of the hotel.

Douglas Upper Douglas Cable Tramway 1896–1929

The Upper Douglas Cable Tramway was built to serve the transport needs of the residents and business owners of Upper Douglas. The Act to build the line was passed on 3 August 1895. This was for a crescent-shaped route commencing at the junction of Victoria Street and Loch Promenade, and then travelling up Victoria Street, Prospect Hill, Bucks Road, Woodbourne Road, York Road and Ballaquayle Road before terminating at the foot of Broadway, where it met the promenade again. The line was to be built by the Isle of Man Tramways & Electric Power Company (who owned the horse tramway along the promenade, which they hoped to electrify) to a gauge of 3 feet and the Act specified 'a tramway operated by wire ropes, cables of chains with a fixed engine'. A continuous cable of 3 miles in length drove the tramway, with grippers attaching the tramcars to the cable. Twelve cars (eight open and four closed) were built by G. F. Milnes of Birkenhead and they each had cross-bench seating for thirty-eight people. The car shed and engine house were located off York Road, and following a satisfactory inspection, the tramway was officially opened on Saturday 15 August 1896, with three cars providing a ten-minute service.

Douglas cable tram No. 75 waiting to commence its journey from Victoria Street c. 1905.

The tramway proved initially popular and in September 1901 was acquired, along with the horse tramway, by Douglas Corporation for £50,000 after the Isle of Man Tramways & Electric Power Company went into liquidation following the collapse of their financier, the Dumbell Bank, in February 1900. The corporation decided to close the latterly little-used Broadway section of the line in 1902, cutting it back to Ballaquayle Road just beyond the depot. Three years later, a connecting track was laid to the horse tramway, and between 1907 and 1911 four new cars arrived. Passenger numbers on the cable tramway remained buoyant up the commencement of the 1920s, but during the winter of 1921/2 Tilling Stevens petrol-electric buses provided the service over the route and thereafter the trams only ran during the summer months before running for the last time on Sunday 18 August 1929.

The distinctive clanking noise of the trams was to be heard no more, but two of the cars (Nos 72 and 73) survived in use as a bungalow and were rebuilt into one fully restored car by the Douglas Cable Car Group in the former York Road depot. The tram bears the number 72 at one end and 73 at the other and is based at Jurby Transport Museum on the island.

Millbrook House, 9, Drury Terrace, Broadway.   Douglas, Isle of Man.
Mrs Talbot

An Upper Douglas Cable Tramway car passes Millbrook House in Broadway on a section of the line that was closed in 1902.

Bucks Road, Douglas. I.o.M.

Car No. 76 of the Upper Douglas Tramway descends Bucks Road *c.* 1905.

Douglas Head Incline Tramway 1900–53

This second cliff railway to be built in Douglas was opened in July 1900 by the Douglas Head Incline Railway Company, formed with a capital of £10,000. The line was built to connect Douglas Harbour with the popular tourist attraction of Douglas Head (with its pierrot concerts, camera obscura and revolving tower) and the Douglas Southern Electric Tramway (1896–1939), which took its passengers on a breathtaking ride along the sea cliffs to Port Soderick. The Douglas Head Tramway had a double track of 4 foot gauge and ran for 450 feet (137 metres) on a gradient of 1 in 4.5. Unusually, it had a bend a third of the way up. The lower station was situated close to the Battery Pier, while the Upper Station, housing the oil engine that powered the line, was seaward of the Douglas Southern Electric Tramway. The two toast-rack cars, with stepped seating, were built by Hurst, Nelson & Company of Motherwell.

The tramway was well-used before the First World War, when the attractions of Douglas Head were at their most popular. The line was closed during the war, but was reopened and on 7 July 1922 it changed hands. During the Second World War, the tramway was closed again, but did reopen in 1949. However, by this time it faced stiff competition from buses and was permanently closed in 1953. In October and November 1955, the line was dismantled and some of the rail was re-used on the Manx Electric Railway on the outskirts of Ramsey. The course of the Douglas Head Incline Tramway is now hard to trace.

Dare's Minstrels entertain the crowds outside the Douglas Head Hotel *c.* 1910.

Left: A view looking down the track of the Douglas Head Incline Tramway showing the noticeable bend a third of the way up.

Below: A toast-rack car of the Douglas Head Incline Tramway stands at the upper station. Note the footpath that runs parallel to the line.

Douglas Cunningham Camp Escalator 1923–68

Liverpool baker Joseph Cunningham and wife Elizabeth founded their first summer camp on the Isle of Man at Laxey in 1892. Two years later, they established their own, all male, camp at Howstrake, before opening a camp at Douglas in 1904, where around 1,500 tents (housing up to eight people each) and a large dining pavilion were erected.

To save the long climb up to the camp from Douglas Promenade, a unique escalator was constructed in 1923. A castle-like entrance building had been built in Switzerland Road in 1914, and from there the escalator ran uphill through a long wooden building to the camp by means of a continuous chain powered by an electric motor. Attached to the chain were individual seats, upon which the passengers sat sideways. A second escalator was added in 1938. J. Skillicorn of Onchan built the entrance building and second escalator and may have built the first one as well. Both escalators comprised of 100 chains, of which forty-eight were visible at one time. The remaining fifty-two chains made their way upside down underneath the escalator before returning to the surface. Campers rode free of charge, but for others it was a penny a ride. The motors for both sets were housed in a concrete building under the top platform.

The 1938 escalator did not run after the 1964 season and the other set was closed four years later. The gateway survives, although bricked-up, and inside the escalators lie derelict.

The tents and dining pavilion of the Cunningham's Young Men's Holiday Camp, Douglas, *c.* 1905.

"ENTRANCE" TO THE CUNNINGHAM CAMP

Above: The entrance to the Cunningham Camp Escalator in Switzerland Road, Douglas, in the 1920s.

Left: A recent view of one of the disused Cunningham Camp Escalators.

Douglas Falcon Cliff Hotel Lift 2 1927–90

Thirty years after the original Falcon Cliff lift had been removed, the owners of the hotel decided that a lift was indeed a convenient way for its patrons to visit the seafront below. The new lift was built on a different site to its predecessor and consisted of a 5 foot gauge single track, 129 feet (39 metres) in length, running in channel section rails set at a gradient of 1 in 1.15. An electric 400 V DC 6 hp motor powered the single car, which could carry six passengers plus an attendant. The balance weight (the weight of the car plus three passengers) ran up and down under the car. The line was constructed by Messrs William Wadsworth of Bolton.

In 1932 the lift was rebuilt, and in 1945, along with the hotel, was sold to brewers Okell & Sons Limited. They converted the electric operation of the lift in 1950 to 415 V three-phase AC power supply. The lift was originally opened all year round, but became seasonal from June to September, and by the time of its closure it ran only on Friday evenings. The end for the lift came with the selling of the hotel in 1990 by Okells. Three years later, the hotel was converted into offices, but the lift was left to decay in situ and its site is now overgrown.

Left: The second Falcon Cliff Hotel Lift at Douglas, which was installed in 1927. A sign advertises that putting and tennis were available in the hotel grounds.

Right: The Falcon Cliff Hotel Lift, photographed in 1990 – its last year of operation.

Folkestone Leas Lifts 1885 and 1890–1966

Folkestone was the first south coast resort to acquire a cliff lift and the third in England after Scarborough and Saltburn-by-the-Sea. The Folkestone Pier & Lift Company promoted a cliff lift in 1883 to connect the famous Leas cliff-top promenade to the beach, but although they eventually built and opened the Victoria Pier in 1888, the lift was built by Folkestone Lift Company, led by local estate agent John Sherwood. John Collins was engaged as the consulting engineer, and John Newman was to carry out the building work, including the attractive lower station designed by Reginald Pope. The upper station, sited just below the Leas, consisted of just a hut for the brakeman. The lift equipment was supplied by the specialist firm of R. Waygood & Company and was to be powered by water. A track gauge of 5 feet 10 inches was decided upon, and two cars, each seating fifteen people, were to run for 164 feet (49 metres) on a gradient of 1 in 1.58. The total cost of the work was £3,224.

The lift was opened on 16 September 1885 in time for the annual regatta when 2,389 passengers were carried. A regular service was commenced five days later and proved to be such a success that an adjoining lift was added in 1890. This ran on a steeper incline of 1 in 1.49, which led to the cars having stepped seating (for sixteen passengers) and end entry doors, as opposed to the side entry doors of the 1885 cars. The track gauge was different too, 5 foot, and the track length was 155 feet (47 metres). At the same time, the lower station was extended to accommodate gas engines and pumps to allow the water to be stored and recycled, rather than to be expensively dumped onto the beach after each journey, as was the case before. Storage tanks were placed under the footway of the Leas by the upper station and under the roadway by the lower station. The cable balance wheels for the two lifts were anchored to a girder buried 20 feet into the Leas. The cost of the 1890 work was £4,262. A second set of tanks were added in 1899 to increase the amount of water that could be stored.

The two lifts often ran to full capacity during the golden age of 'Fashionable Folkestone' up to the start of the First World War in 1914. The Leas was one of the finest marine promenades in the country, while the seafront had the Victoria Pier, switchback railway, bathing establishment and concert parties. Improvements to the lift's pumping machinery was carried out in 1909 when they were converted to electric working, and in 1921 when the Crossley gas engines were replaced by electric motors. The lifts remained well used between the wars, but were closed in the Second World War when the lower station was used as a Home Guard post. They were reopened in 1948 with new electric motors to drive the pumps.

By the 1960s, Folkestone's heyday as a resort was over and the Leas Lifts needed major improvements. The 1890 lift carried its last passengers in October 1966 and the Folkestone Lift Company went into voluntary liquidation the following year. Folkestone Borough Council took over the running of the 1885 lift and added a mid-track braking system in case of an emergency before they were absorbed into Shepway District Council in 1974. In 1992 the council fitted safety measures to ensure that the car doors had to be closed before they could move, but in 2009 they announced that they could no longer afford to run the lift. The lease was surrendered back to the Folkestone Estate, and they carried out a refurbishment of the lift before a licence was granted back to the Folkestone Leas Lift Community Interest Company to operate it. The lift was reopened in July 2010. The redundant tracks of the 1890 lift were removed in 2013, although one of the stepped cars was restored and is on display at the Elham Valley Line Trust Countryside Centre & Railway Museum.

The future of the Leas Lift was once again thrown into doubt when, on 17 October 2016, the Health and Safety Executive issued an improvement notice regarding the braking system, which was said to be unreliable and prone to failure. The Folkestone Leas Lift Company could not meet the estimated £80,000 cost to replace the braking system and surrendered its operating licence on 27 January 2017. Fortunately, a new community interest company was established in October 2017 with the intention of reopening this historic cliff lift, which is Grade II listed and the second oldest remaining water-balance lift in Britain.

Above left: The 1885 Leas Lift at Folkestone, photographed from the Victoria Pier *c.* 1889 before the adjoining 1890 lift was added.

Above right: A postcard from *c.* 1903 showing both the 1885 and 1890 Folkestone Leas Lifts in running.

The upper station of the Folkestone Leas Lift *c.* 1910, with both types of lift cars visible.

The 1885 Folkestone Leas Lift in operation *c.* 2000 with the disused track of the 1890 lift to the right of it.

Folkestone Sandgate Hill Lift 1893–1918

Following the success of the Leas Lifts, the Sandgate Hill Lift Company was formed in 1890 with a capital of £6,000 to provide a lift from the extreme western end of the Leas down to the neighbouring village of Sandgate, where a horse tram (opened in 1891/2) could be caught to Hythe. Like the Leas Lifts, the design of the stations was placed in the hands of local architect Reginald Pope, who was assisted by C. E. Robinson. John Newman prepared the track bed and Waygood provided the lift equipment, which was worked on the water-balance principle. A 17,000-gallon water tank was placed under the upper station on the Leas, which had a waiting room and an unusual railway-type semaphore signal with the arm displaying the word 'LIFT' when the lift was open. From the upper station, the 5 foot 6 inch gauge double track ran for 670 feet (204 metres) and crossed Radnor Cliff Crescent by way of an ornate bridge, but due to the changes in gradient of the undulating route from 1 in 4.75 to 1 in 7.04, the cars had to be independently operated with a brakeman. He operated the sliding brakes, which could be applied automatically if the speed of the cars exceeded the limit allowed. The journey time was eighty seconds, with the cars passing each other just on the upper side of the bridge. The undercarriage of the cars, with the system of patent braking, were constructed by Messrs Jones Bros of Lynton and the bodies were made locally by Messrs Worthington Bros of Hythe. They could each seat sixteen passengers and provision were made to carry a bath chair in one of them (there was a 6d charge for bath chairs). Access to the cars was by sliding doors. The lower station was situated on Upper Folkestone Road (now Sandgate Hill) and had a 21,000-gallon water tank and a gas engine, which pumped the water back up to the upper tank. For the comfort of passengers, there was also a waiting room, ladies' cloakroom and a lavatory.

The lift was officially opened on Monday 20 February 1893, but proved not to be as profitable as the Leas Lifts. In 1901, author H. G. Wells built a house adjoining the course of the lift and was an interested observer of it as the cars clanked past his front door.

The coming of the First World War eroded the precarious finances of the Sandgate Hill Lift Company still further and in July 1918 they took the decision to suspend the service. During a meeting of the company directors in April 1919, it was revealed that £500 was needed for repairs to the lift and the decision was taken to close it. An offer by Sandgate Council to buy the lift for £500 was turned down, and eventually, on 25 June 1923, it was acquired for £650 by London architect A. M. Cawthorn, who dismantled it. The Sandgate Hill Lift Company was formally wound-up in April 1924.

The site of the upper station on the Leas is now covered by a house called 'Sandgate Point', but the lower station still survives, albeit in a much-altered state, as 'Croft House'. The abutments of the bridge over Radnor Cliff Crescent are still evident.

The Sandgate Hill Lift seen from the Folkestone Leas on a postcard sent in 1903. The coastal village of Sandgate can be seen below.

An attractive postcard of the Sandgate Hill Lift and its attractive lower station in 1904. The building above the station is Spade House, built for author H. G. Wells.

Folkestone Metropole Lift 1904–40

The western end of Folkestone's Leas acquired its own lift on 31 March 1904 with the opening of the Metropole, or West Leas, Lift. This lift was promoted by the Folkestone Metropole Lift Company; formed by directors of the Folkestone Lift Company and registered on 12 May 1903 with a capital of £7,050 in £5 shares. The 5 foot 6 inch gauge double track ran 96 feet (29 metres) down the cliff face opposite the Metropole and Grand hotels to the Lower Sandgate Road, where there was an attractive red brick entrance building designed by Reginald Pope. In addition to having a waiting room, the building housed a pair of gas-powered Crossley pumps, which pumped seawater to the water storage chamber built into the Leas. The upper station consisted of just a small brakeman's hut. The two cars had seating for sixteen people and sliding end doors. As was the case with the Leas Lifts, resident engineer John Collins built the lift and oversaw its construction in association with R. Waygood & Company.

Although never as busy as its cousin at the other end of the Leas, the Metropole Lift provided a useful and convenient service for Folkestone's visitors until its closure in August 1940, when the town was very much in the firing line during the Battle of Britain. Sadly, the lift fell into disrepair and was never reopened. The operating company went into liquidation in June 1951 and the lift was dismantled. Little trace can be seen of existence, although the site of the water storage tank on the Leas is identifiable.

A postcard view of the attractive Folkestone Metropole Lift soon after it was opened in 1904. The lift attendant can be seen standing outside the lower station.

The upper station of the Folkestone Metropole Lift seen on a postcard *c.* 1905.

The lift attendant awaits the arrival of two customers for the Folkestone Metropole Lift *c.* 1910.

Hastings West Hill Cliff Railway 1891

Hastings has two cliff lifts, of which the oldest is the West Hill Cliff Railway. The lift interestingly runs in a 363-foot-long tunnel through a natural cave for much of its length and provides easy access to Hastings Castle and the Smugglers Adventure in St Clement's Caves.

The lift was promoted by the Hastings Lift Company with a £10,000 budget and Messrs A. H. Holme and C. W. King commenced work on it in January 1889 to a design by F&S Plowman. Objections to the plans led to expensive delays, which pushed the final cost up to £16,000. The lift was opened on 25 March 1891 and featured an imitation Mansfield stone façade on the George Street lower station erected by Elliot's Patent Stone Company. The 6 foot gauge track ran for 500 feet (152 metres) on a gradient of 1 in 2.9 and a vertical rise of nearly 170 feet, and the two cars (built by the Midland Railway Carriage & Wagon Company of Birmingham) could hold sixteen passengers each. Waygood supplied the lift winding gear, which was powered by a 40 hp Crossley gas engine hidden in an underground engine room at the upper station. The gearing attached to the steel ropes drove in either direction to ensure that the cars were under control and could be halted at any time. There were additional braking safety features.

The lift attendant waits for any more passengers on the West Hill Cliff Railway *c.* 1900.

In 1894, the Hastings Lift Company went into liquidation due to the cost of constructing the lift and it passed to the Hastings Passenger Lift Company. They held it until 1947, when it was acquired by Hastings Borough Council for £4,500. In 1971 they replaced the diesel engine installed in 1924 with electrical working installed by the British Ropeway Engineering Company. To celebrate its centenary, the lift was fully refurbished in 1991, and further improvements were carried out between 2003 and 2005, when the track and sleepers were replaced.

The West Hill Cliff Railway is a well-used attraction and is open all year round.

Above left: A postcard of the Hastings West Hill Cliff Railway *c.* 1905 showing the tunnel through which it runs.

Above right: The George Street entrance into the West Hill Cliff Railway, Hastings, photographed in September 2017.

Hastings East Hill Cliff Railway 1903

Hastings' other cliff lift, the East Hill Cliff Railway, is the steepest funicular railway in Britain with a gradient of 1 in 1.28. The builders of the West Hill Cliff Railway, the Hastings Lift Company, wished also to provide a lift on the East Hill, but their financial troubles scuppered their plans. However, Hastings Council, who had spent £11,000 on purchasing the East Hill for public recreation, then decided to build the lift themselves. A design was produced by the Borough Engineer, Philip H. Palmer, who recommended a water-balance lift as the water supply could be gained easily from the nearby refuse destructor.

The work of constructing the lift was given to Messrs Easton of Erith Ironworks and it was opened on 16 April 1903, having cost £6,000. The double track was of 5 foot gauge and ran for 267 feet (81 metres). The two cars seated twenty passengers each and had 600-gallon tanks between the underframes. Up on the East Hill, the upper station resembled a miniature castle and housed 1,200-gallon tanks in each of the towers. The lower station on Rock-a-Nore Road was built in the style of a small cottage and was erected over a 23-foot-deep chamber housing the cross-over pulley for the cable. There was also the discharge tank for the water once the descending car had reached the bottom. The water was then electrically pumped back up to the tanks at the top. Four steel wire ropes connected the two cars, which were supported on hardwood rollers in the centre of each track, and there were two safety braking systems.

Looking down from the upper station of the East Hill Cliff Railway towards the fishing quarter on the beach in the 1920s.

The lift was a reasonably successful concern for the council, and in the mid-1970s they spent £35,000 on converting it to electrical working. An additional £6,900 was spent on two new cars, each seating sixteen passengers. However, in June 2007 the lift was closed after a fault in the control panel caused the cars to fail to crash into the stations. A major refurbishment resulted in new control systems and cars and the lift was reopened in March 2010.

The East Hill Cliff Railway is normally open between April and October and provides access to Hastings Country Park.

Two views of the East Hill Railway, Hastings, taken a century apart: one *c.* 1910 and the other in May 2010.

Laxey Browside Tramway
1890–1906

The Isle of Man's most obscure cliff lift was the Laxey Browside Tramway, which took visitors up to see the famous Laxey Wheel (also known as 'Lady Isabella'), built in 1854 to pump water from the Great Laxey Mines. The tramway was built by members of the mining family and was opened on 16 August 1890. The 6 foot gauge double track line ran for 390 feet (118 metres) on a gradient of 1 in 4 and the water used to power it was piped from Lady Isabella's tail rack that was also used to power the mighty wheel. An annual rent of £25 was paid to the mining company for the water used, and the task of emptying the tank of the ascending car was carried out by a young bung boy. The two toast-rack cars had eight fixed benches, which could seat twenty passengers each. Men had to pay a penny to ride upon the lift and a ha'penny down, but women and children apparently travelled free.

In 1893, one of the cars crashed into the bottom station when a brake was not correctly applied. The other car jammed itself into the top station. No passengers were hurt, but John Callow, the mining company clerk, was hit by the ascending car and suffered minor injuries. In 1906, the line was closed after the mining company refused to supply the water needed to power the line. This action followed a disagreement about how much the tramway company should pay for the water used. The line was sold to Robert Kelly of Douglas, and by 1910 had been dismantled. There is very little evidence today of its existence.

Below left and right: Two grainy, but rare, views of the Laxey Browside Tramway. One shows the upper station with the Laxey Wheel behind, and the other is looking down the track from the station.

Llandudno Great Orme Tramway 1902

The Great Orme Tramway is Britain's only remaining cable-hauled tramway and is, in reality, a double reverse funicular with two cable-hauled cars plying each of the two sections of the line. The lower section runs for 2,624 feet (800 metres) from the Victoria station through narrow streets to Halfway station, where the cable-winding and engine house is located. The upper section of the tramway runs for 2,460 feet (750 metres) from the Halfway station to the summit, where there is a visitor centre for the Great Orme Country Park. Both sections are operated by a winding engine powering two large cable drums: one for the ascending car and the other for its descending twin. On the upper section, the cable runs exposed between the rails over rollers and pulleys. There are three cables on this section: one linking the two cars via a large sheave behind the Summit station, and the other two linking the two cars to the winding drums. The cable on the lower section is buried beneath the double track, which is of 3 foot 6 inch gauge. The four tramcars, Nos 4–7, are survivors of the original seven cars built by Hurst, Nelson & Company of Motherwell. Nos 1–3 were freight vans withdrawn in 1911. The total length of the line is 1 mile and it climbs 679 feet on a maximum gradient of 1 in 3.9.

The Great Orme (Y Gogarth in Welsh) rises on three sides above the town of Llandudno, which was developed as a select seaside resort by the Mostyn family from the 1840s. There had been various proposals on how to transport visitors up to the summit of the Great Orme and a cable tramway was proposed by the Great Orme Tramway Company, which was formed with a capital of £25,000 in £5 shares. The Great Orme Tramways Act was passed in 1898 and the construction of the line was commenced in April 1901 with Richard White & Son of Widnes supplying the permanent way materials, winding gear and rolling stock, and Thomas and John Owen of Llandudno laying out the track route. The lower section of the line was opened on 31 July 1902 and the upper section on 8 July 1903. In 1904, the Victoria station at the town terminus was built on the site of the Victoria Hotel.

The tramway was an immediate success and improvements to the line included the purchase of a new engine in 1914 to work the lower section, and a new boiler house and large boiler in the late 1920s. Unfortunately, on 21 August 1932, a tramcar on the lower section broke loose from its cable due to a broken drawbar, became derailed and crashed into a stone wall. The driver and a twelve-year-old girl were killed, and several passengers injured. The operating company went into liquidation and the line was sold to the Great Orme Railway Company. New safety measures were installed, and the tramway was reopened on 17 May 1934.

On 1 January 1949, Llandudno Urban District Council took over the tramway and in 1957 its motive power was converted from steam to electric working by the English Electric Company. In 2002, a Doppelmayr inductive loop system was installed where signals are relayed between the winchman and drivers via cables at the trackside and sensors mounted below the tramcars. A track position indicator on the control panel in the winding house shows the position of the cars, and the cables are marked to show where the cars should stop at their termini. Another major improvement to the line was completed in 2001 with the building of a new Halfway station and engine house as part of a phased £4 million programme to refurbish the tramway. An exhibition on the history of the tramway was included in the new station.

The Great Orme Tramway remains one of the major tourist attractions of North Wales and a round trip on it takes fifty minutes. The line is normally open from March to October.

A car on the Great Orme Tramway halts for the postcard photographer on its climb up to the summit. The postcard was sent in 1915.

Great Orme Tramway Cars Nos 6 and 7 on the section of the line between the Halfway station and the summit in the 1990s.

A photograph taken in 1996 showing Great Orme Tramway No. 4 descending to the Victoria station.

Great Orme Tramway No. 5 at the Victoria station in 2007. Car Nos 4 and 5 operate the lower section of the line between Victoria and Halfway stations.

Lynton and Lynmouth Cliff Railway 1890

One of the most popular and widely used cliff railways is the Grade II listed water-balance lift that connects the twin north Devon villages of Lynton and Lynmouth. At 862 feet (262 metres), it is also the longest of the 'traditional' seaside cliff lifts and was the venture that brought George Newnes and George Croydon Marks together in partnership.

Before the arrival of the cliff railway, the only way to travel between the picturesque coastal village of Lynmouth and Lynton 600 feet above was by a steep and tortuous road. In 1885, two prominent residents, London lawyer Thomas Hewitt, and John Heywood, Chairman of the Lynton Local Board, promoted a scheme to build an esplanade and pier in Lynmouth, with a lift to connect them with Lynton. The Lynmouth Promenade Pier & Lift Provisional Order received official sanction in October 1886 and the promenade was duly constructed by Jones Brothers of Lynton. It was Bob Jones who recommended to Hewitt and Heywood that George Croydon Marks should engineer the cliff railway, and, after visiting him, Hewitt's friend George Newnes agreed to put up most of the finance to build it. Work began on blasting away the cliff face in late 1887, and in June 1888 Newnes, Jones and Hewitt filed a patent for four separate braking systems for the railway, so designed that they could be independently operated from each other. A further Act of Parliament, passed for the Lynmouth & Lynton Lift Company, was obtained in 1888 to help override local objections to the scheme.

The Lynton & Lynmouth Cliff Railway was formally opened by Mrs Jeune, Lady of the Manor of Lynton, on Easter Monday 9 April 1890. Newnes was so enchanted with the area, termed the 'English Switzerland', that he built a house there. The pier was never built (perhaps because Newnes feared his beloved twin villages would be swamped with trippers arriving from the Bristol Channel steamers), but Newnes did help finance the building of a town hall for Lynton and the remarkable Lynton & Barnstaple narrow gauge railway.

The cliff railway was constructed by Bob Jones and rises a vertical height of 500 feet on a gradient of 1 in 1.75. The two tracks have a gauge of 3 feet 9 inches and are just 8 inches apart, except at the passing point. Unusually, there was an intermediate station until 1923 where the cars passed beneath the North Walk Bridge. The original cars could be taken off the undercarriage, which could then be used for transporting goods, or even motor cars, but they were replaced in 1947. The storage tanks under the cars can each hold 700 gallons of water, which is taken from the West Lyn River and piped by gravity for 1 mile to the storage tank at the upper station. Up to 60,000 gallons of water per day can be drawn from the river. The cars are controlled by driver/conductors (standing on the outside platforms) who release the brakes, and the water from the car storage tank of the descending car once it has completed its journey. The speed of the cars is controlled by a governor actuating a hydraulic slipper brake. There are additional braking systems using piston-operated callipers to grip the crown of the rails, and wedges and brake blocks, which can be used in an emergency. Two wire hauling cables with a braking strain of 27.5 tons are attached to each car, suspended by a double vee pulley wheel 5 foot 6 inches in diameter, while two tail ropes are attached to balance the weight of the 10 cwt upper ropes and to steady the motion.

The lift is still owned by its original operating company, who also run a café at the upper station. Journey time is two and a half minutes and it is normally open between February and November.

Lynmouth. The Cliff Railway.

The Cliff-Railway, Lynmouth

Above left: The inaugural run of the Lynton and Lynmouth Cliff Railway on 7 April 1890, with George Newnes on the outside platform, was used for this postcard, issued *c.* 1902.

Above right: The Lynton and Lynmouth Cliff Railway seen in 1904, with two of the staff posing for the postcard photographer.

Two views of the Lynton and Lynmouth Cliff Railway taken in August 2017. The top photograph shows the car passing the site of the former intermediate station at North Walk.

Machynlleth Centre for Alternative Technology 1992

The Centre for Alternative Technology is an education and visitor centre demonstrating practical solutions for sustainability and for green living. It was founded in 1973 on the site of the disused Llwyngwern slate quarry near Machynlleth in Mid Wales and was opened to the public in 1975. The centre has grown to become Europe's leading eco centre.

To ease the climb from the car park to the centre, a cliff lift was opened on 7 July 1992, which, in the interests of energy conservation, runs on the water-balance principle. The water to operate the lift is drawn from a lake. The line has two parallel tracks of 5 foot 6 inch gauge and runs for 197 feet (60 metres) on a gradient of 1 in 1.81. The two totally enclosed cars, built by the Lancastrian Carriage & Wagon Company of Heysham, have stepped seating for seventeen passengers each, although this is usually restricted to sixteen in the descending car. Wheelchairs, motor scooters and pushchairs can also be carried. The two station buildings are constructed of wood, with the control room for the lift being housed in the upper station. The whole system is computer controlled and has a five-stage braking process. The lift has a journey time of one minute and forty-five seconds and is normally open between Easter and the end of October.

Above left and right: Two views of the cliff lift at the Centre for Alternative Technology near Machynlleth, opened in 1992.

Margate Cliftonville Lido Lift 1912–72

The tiny single car lift at the Cliftonville Lido ran for only 69 feet (21 metres) from the bathing pool to the cliff top on a gradient of 1 in 1.38 and was unusually placed parallel to the sea. The lift was installed in 1912 by R. Waygood and was operated by them through their subsidiary company Cliff Lifts Limited. The track gauge was 5 feet and the car could carry twelve passengers. The counterbalance weight was placed in a vertical shaft.

The lift was opened adjacent to the Clifton Baths, which dated back to 1824, and which, as well as housing indoor baths of various varieties, also had bathing boats and machines operating from the shore below. Other facilities included a library, billiard and music rooms and a terraced walk. In 1919, John Iles and his Margate Estates Company, who had acquired Margate's Hall by the Sea and were transforming it into the famous Dreamland amusement park, purchased the Clifton Baths site and erected a concert hall on the site of the baths. In 1927, a large open-air bathing pool (lido) was built out to sea, and other attractions added included a café, bars, bandstand, zoo and aquarium.

The lift was usually open only between June and September, but was closed in 1972. The track bed and cable wheel are still visible. The bathing pool was filled in during 1977 and the theatre was demolished, yet some of the lido complex remains, including the distinctive sign.

The Cliftonville Lido Lift at Margate in the 1960s, which, unusually, ran parallel to the sea.

The Cliftonville Lido complex in the 1960s. The lift was sited parallel to the path coming down from the cliff top.

The Cliftonville Lido Lift, photographed *c.* 1970.

Margate Cliftonville Walpole Bay Lift 1934–2009

The Walpole Bay lift was opened in June 1934 by Margate Borough Council to provide access from Queens Promenade to the open-air tidal pool below. The lift has concrete shaft 30 feet in height and 12 feet wide at the base, tapering towards the top, which has a stepped cornice. The single electrically operated elevator car could carry fifteen passengers and in the 1950s the fare was 3*d* for adults and 2*d* for children.

In 1998, the lift was refurbished and was open from June to September, when it was free to use. It was given a Grade II listing for its art deco design with Egyptian influences. The Walpole Bay Tidal Pool, dating from 1900, is also Grade II listed. However, in 2009, the lift was closed by Thanet District Council due to lack of usage and it currently remains out of use.

Above left and right: Two views of the closed Cliftonville Walpole Bay Lift at Margate, photographed in September 2017.

Matlock Cable Tramway 1893–1927

The thermal springs of Matlock Bath had been popular with visitors since the eighteenth century, resulting in Matlock becoming a leading inland resort. In 1853, John Smedley opened his hydropathic establishment on Matlock Bank and, following the arrival of the Midland Railway in 1863 and the opening up of Matlock to visitors from the north, other hydros soon opened in the area, all specialising in water-based systems to revitalise health. However, it was a long climb up to the Matlock Bank hydros from the railway station at Matlock Bridge, and Job Smith proposed the laying of a cable tramway on the San Francisco model to connect them. The local board did not want to build it, but in 1890 George Croydon Marks, fresh from engineering the Lynton & Lynmouth Cliff Railway, visited Matlock and heard about the proposed tramway. He told George Newnes about it and Newnes, who was born at Matlock Bath, agreed to help finance the building of the line. Newnes and Smith formed the Matlock Cable Tramway Company with a capital of £20,000, and Croydon Marks was appointed engineer. However, due to difficulties with the steep gradient of Bank Road, Croydon Marks consulted a specialist on cable tramways, William Newby Colam, who was engineer and consultant of other cable tramways in the UK. Colam and Croydon Marks prepared the plans for the tramway and appointed contractors Messrs Dick, Kerr & Company of Kilmarnock. The tramcar depot, housing the engine and boiler houses, was placed at the upper terminus at the top of Bank Street, at the corner of Wellington Street and Rutland Street. The contract to build the depot, and its 100-foot-high chimney, was let to Messrs W. Knowles & Sons to plans prepared by J. Turner of Matlock Bridge and work commenced in the summer of 1891. The line itself was to be single track, of 3 foot 6 inch gauge, with a passing place at Smedley Street. It was to run for half a mile, from Crown Square up Bank Road to the depot, and rise 300 feet with a steepest gradient of 1 in 6.5. Several tight curves would have to be negotiated. The steel cable was carried beneath the surface of the road through a cast iron tube, with the conduit slot rails laid in the centre of the two rails through which the gripper from the cars connected with the cable. At intervals along the route, guiding pulleys were carried in the sub-channel for supporting the cable, which travelled at the rate of 5½ mph in an endless journey between the large pulley and guiding wheels in the depot and at the terminal wheel placed in a chamber beneath Crown Square before returning in the same tube, the two runs being kept apart from touching by a series of small pulleys. To bring the cars to a halt, the driver released the gripper from the cable and brought the cars to a standstill by means of a brake. In addition, there was also an emergency brake. Three tramcars, to carry thirteen passengers inside and eighteen on the open top deck, were ordered from G. F. Milnes & Company of Birkenhead. A fourth car, purchased in 1911 from the redundant Birmingham cable line and remodelled as a single-deck vehicle, was later added.

The tramway was opened on 28 March 1893 and was at the time the steepest in the world on a public road. Many of its customers were visitors to the hydros, who paid a *2d* fare to go up Bank Road but only a penny to go down. In June 1898, George Newnes bought out his fellow shareholders in the Matlock Cable Tramway Company for £20,000 and presented the tramway to Matlock District Council. They added a very ornate tram shelter and clock tower in Crown Square in 1899 but the tramway was to prove a financial burden, largely due to high maintenance costs. By the early 1920s, the council were facing deficits of well over £1,000 a year on the tramway and on 27 September 1927 it was closed.

Reminders of the cable tramway can be seen in the surviving depot, which is currently in use as a car repair centre, and part of the Crown Square shelter, which was placed in Hall Leys Park after closure.

A Matlock cable tram can be seen at its lower terminus in Crown Square on this postcard sent in 1910.

From Crown Square, the Matlock cable line ran up Bank Road to the top of the hill. A tram can be seen on the hill on this postcard sent in 1928, one year after the tramway closed.

Port Soderick Cliff Lift
1898–1939

Situated south of Douglas, Port Soderick was developed by the Forrester brothers as a tourist destination in the latter part of the nineteenth century. There were natural caves off the beach to explore, and a hotel, pavilion and restaurant were built. In 1896, the Douglas Southern Electric Tramway provided a station at the top of the cliff, but its passengers had to navigate 150 steps down to the beach and attractions.

To make life easier for their patrons, the Forresters acquired the defunct Falcon Cliff Hotel Lift at Douglas and re-erected it at Port Soderick in 1898. The reconstructed lift was longer than it was at Falcon Cliff, but retained the same 4 foot gauge, and ran down the incline on a timber viaduct supported by stone pillars. New cars were built, which

A close-up view of the cliff lift at Port Soderick, showing the pagoda hut that served as the lower station and the timber viaduct upon which the line ran. A set of swing boats can be seen on the right.

were sloped to match the line's gradient; the original cars were converted into kiosks. An oil engine powered the lift. The upper station, situated close to the terminus of the Douglas Southern Electric Tramway, was constructed of corrugated iron, while the lower station had a pagoda hut. Approval for the lift to carry passengers was granted on 11 July 1898.

The lift proved to be a useful asset for those visiting Port Soderick, but upon the outbreak of the Second World War in September 1939, both the lift and the Douglas Southern Electric Tramway were closed, never to reopen. The lift was dismantled in 1949 and the two cars were used as hen houses until, apparently, a herd of cows were poisoned from chewing the lead paint on the roofs. The site of the lift is now overgrown, although some of the stone foundation pillars of the viaduct can still be seen.

The attractions of Port Soderick can be seen in this postcard from 1904. In addition to the cliff lift, they include a pavilion, restaurant, hotel and camera obscura.

Port Soderic.

A postcard of the Douglas Southern Electric Tramway, which gave its passengers a spectacular scenic ride from Douglas to Port Soderick, where the cliff lift took them down to the beach.

Ramsgate Marina Lift
1908–70

The Kentish resort of Ramsgate had three elevator lifts, of which only one, the Harbour Lift, still operates. The Marina, or East Cliff, Lift was the first to be opened, on 31 May 1908 by Cliff Lifts Limited, a subsidiary company of Waygood, who provided the electric motor and lift equipment. The lift, which was designed by G. Graham Tucker of Ramsgate and erected by local firm Grummant, could carry fifteen people and was housed in a 68-foot (20-metre) brick-lined shaft.

On 31 March 1919, both the Marina and Harbour lifts were acquired by Ramsgate Corporation for £2,000. However, in January 1970, the Marina Lift was demolished, although an enclosed service ladder remained in place for a time. Now, there is little evidence the lift ever existed.

A postcard issued in 1912 by Cliff Lifts Limited advertising their Marina Lift at Ramsgate. They also operated the Harbour Lift at Ramsgate, the Broadstairs Viking Bay Lift and Cliftonville Lido Lift at Margate.

Another Cliff Lifts Limited advertising card of the Marina Lift, Ramsgate, this time showing the ornate upper entrance.

Following the demolition of the Marina Lift in 1970, this enclosed service ladder was put in place and is seen here in May 1981. This also no longer exists.

Ramsgate Harbour Lift 1910

The Ramsgate Harbour Lift was opened by Cliff Lifts Limited in August 1910 and was an attractive structure housed in a 58-foot-high (17-metre) brick tower crowned with a lead dome. Ornate masonry was provided at the upper entrance in Wellington Crescent spelling out the word 'LIFT', and geometric floor tiles lined the upper entrance path and lower hall. The lift car, supplied by Waygood, could carry twenty passengers, although this was subsequently reduced to sixteen when the lift was rebuilt in the 1950s.

In 1919, both the Harbour and Marina Lifts were acquired by Ramsgate Corporation for £2,000. Unlike the Marina Lift, the Harbour Lift escaped demolition, and on 23 April 1999 it was reopened following a £220,000 scheme to restore it to its Edwardian splendour.

The lift is a Grade II listed building and is free to use when open between Easter and September.

A postcard showing the upper section of the Harbour Lift, Ramsgate, shortly after it was opened in 1910. Note the clock inscribed with 'Waygood Lifts'. Waygood built the lift through its subsidiary company, Cliff Lifts Limited.

The upper entrance to the Harbour Lift, Ramsgate, photographed in April 2001, two years after it had been fully restored.

The Harbour Lift, Ramsgate, photographed in February 2001 during the closed winter season.

Ramsgate West Cliff Lift 1929–93

The Western Undercliff at Ramsgate was officially opened by the Prince of Wales on 24 November 1926, and three years later it was provided with an elevator lift by Ramsgate Corporation at a cost of £3,328. Designed by John Burnet & Partners, and built by W. W. Martin, the lift was built in concrete with attractive blue glass tiles running the length of the 78-foot (23-metre) shaft. The single lift car could carry up to twenty passengers.

The expansion of the ferry port led to the isolation of the western beach and in 1993 the lift was closed. Five years later it was given Grade II listed status, and despite remaining unused, it is maintained in good condition.

Above left and right: Two postcard views of the West Cliff Lift at Ramsgate in 1930, a year after it was opened.

Two photographs of the closed, but still well-maintained, West Cliff Lift at Ramsgate in August 2017.

Saltburn-by-the-Sea Cliff Hoist 1870–83

As part of the development of the resort of Saltburn-by-the-Sea on the Cleveland coast of Yorkshire, a promenade pier was opened in May 1869, and to ease access to it from the town above, a vertical wooden hoist was provided by the pier's builder, John Anderson. Opened on 1 July 1870 for the Saltburn Pier Company, the hoist had a single cage that could carry twenty passengers up and down the 120-foot drop and was approached by a narrow walkway at the top. It was worked using the counterbalance principle of using water to fill a tank to descend it to the bottom, and then emptying the tank to allow the cage to go back to the top. Guide ropes helped secure the hoist.

The hoist was a potentially frightening experience for some of its customers, particularly if there was a wind blowing. It was also operatically erratic, and the cage had the disconcerting habit of sticking halfway up. In 1883 the hoist was condemned as unsafe and was removed to make way for a conventional cliff railway.

Above left: A carte de visite showing the Cliff Hoist and pier at Saltburn *c.* 1870. A walk out to the lift cage took some courage to do!

Above right: A postcard sent in 1904 remembering the Cliff Hoist, which was dismantled in 1883 and replaced by a cliff railway.

Saltburn-by-the-Sea
Cliff Lift 1884

To replace the rickety condemned hoist, the owners of the Middlesbrough Estates commissioned George Croydon Marks of the Tangye Engineering Company to design a water-balance incline tramway. The water was to be taken from a spring in the cliff face and to be collected from the descending car to be stored in a 30,000-gallon tank at the lower station. From there it would be pumped up to the 18,500-gallon storage tank at the top using Crossley gas engines. The contract to build the two cars was placed with the Metropolitan Railway Carriage & Wagon Company of Birmingham. They were designed to seat twelve persons each, with an over-body sitting on a triangular sub-frame housing a 350-gallon water tank. The cars had striking stained glass windows, which were removed in 1955 when new car bodies were introduced. Fortunately, the stained glass was reinstated in 1991 on the aluminium-bodied cars introduced in 1979. The track was 207 feet (63 metres) in length and had a rise of 120 feet on a gradient of 1 in 1.40. The original gauge was 3 feet 9 inches, but this was amended to 4 feet 2 inches in 1921. In addition to housing the pump room, the lower station had the ticket office and waiting room, but the upper station consisted of just a hut for the brakeman.

An Edwardian postcard showing the upper station of the Saltburn Cliff Lift, consisting of just a brakeman's hut.

SALTBURN-BY-THE-SEA, THE INCLINE TRAMWAY

The lift was opened on Monday 2 June 1884 and has remained a popular attraction ever since. It perfectly complements the pier (alas, half the length it once was), which is the only pleasure pier in the north-east of England. The lift has been well maintained and regularly upgraded. The original Crossley gas engines were replaced in 1913 by an electrically operated DC generator and pump, which in turn was replaced by an AC pump in 1924. In 1998, the main winding wheel was replaced for the first time, and a new hydraulic braking system was installed by Skelton Engineering. An extensive refurbishment was carried out between September 2010 and April 2011 at a cost of £30,000, and in 2014 the upper station was restored to its original design. Now owned by Redcar & Cleveland Council, Saltburn is the oldest water-balance lift in Britain. It has a journey time of fifty-five seconds and is open every weekend between March and October, and daily during the peak summer season.

Left and below: Two postcard views from *c.* 1904 showing the Saltburn Cliff Lift as seen from the pier.

Two photographs looking down the Saltburn Cliff Lift towards the pier below, one taken in the 1960s and the other in the 1990s. Note that the pier has been shortened; this was due to storm damage in 1974 and partial demolition of the structure.

Scarborough Spa Cliff Lift 1875

Otherwise known as the South Cliff Tramway, the Spa Cliff Lift was the first cliff lift to be constructed in the United Kingdom. The prime mover in the building of the lift was Mr Hunt of the Prince of Wales Hotel, who wanted easier access for his patrons to the Spa entertainment complex on the foreshore below. In 1873, he helped form the Scarborough South Cliff Tramway Company, which had a capital of £4,500. The lift eventually cost £8,000 upon its completion and opened on 6 July 1875. Built to the designs of William Lucas and constructed by Crossley Brothers of Manchester and Stewart & Bury of Scarborough, the lift runs for 286 feet (87 metres) on a gradient of 1 in 1.75 and has a track gauge of 4 feet 8½ inches. The hydraulic water-balance system was designed was designed by Tangye of Smethwick, Birmingham, and was powered by two Crossley gas engines (soon replaced by steam pumps in 1879), which pumped sea water to the upper station, from where it was transformed to the tank under the cars. The two cars were constructed by the Metropolitan Carriage Company of Birmingham and could each seat fourteen passengers.

A crowd had gathered on the beach to view the newly opened Scarborough South Cliff Tramway in 1875 – the first cliff lift to be opened in Britain.

On the opening day 1,400 passengers used the lift, each paying a penny fare. It proved to be a profitable venture, and was well-used by the patrons of the South Cliff hotels who used it to get back to their lodgings after a late-night show at the Spa. In 1935, two new cars, built by Hudswell Clarke & Company and each seating twelve, were installed when the lift was converted to electric working. Fully automated operation was introduced in 1997, but on 2 September 1998 eleven passengers were injured when the descending car hit the buffers at the lower station. A fault in the electrical braking system was to blame and the lift remained closed until 14 May 1999 while a back-up braking system was installed. In 2007, the lift was closed for nine weeks to allow the track and sleepers to be renewed.

Scarborough Borough Council acquired the lift for £110,000 on 1 November 1993 but it is currently operated by the Scarborough Spa.

Above left: A postcard of the South Cliff Tramway, Scarborough, issued *c.* 1905.

Above right: Now known as the Spa Cliff Lift, the tramway is seen here in September 2015.

Scarborough Queens Parade Tramway 1878–87

Scarborough's second cliff lift was the short-lived and very obscure Queens Parade Tramway in the North Bay. Promoted by the Scarborough Queens Parade Tramway Company with a capital of £3,500, the lift and its machinery were designed by Henry Holt of Leeds and work on constructing the lift was commenced by Messrs Wade & Son of Leeds in 1878. Running from Queens Parade down to the Promenade Pier, this water-balance lift was 218 feet (66 metres) in length and had a gradient of 1 in 2.3. The track gauge was 4 feet and part of the right of way was in a cutting where it ran under a bridge. The approach to the top of the cliff was on an iron girder structure carried on cast iron piles. The lower station had a waiting room and tollhouse, and housed a 3½ hp Otto patent silent gas engine, supplied by Messrs Crossley Brothers of Manchester, which drove a double-acting geared pump set under the station floor that was capable of raising water from a brick tank, which formed the foundations of the station. The pump, supplied by Messrs Hathorn, Davey & Co. of the Sun Foundry, Leeds, was capable of lifting about 12 tons of water per hour into the upper storage tank of 80 capacity. The upper station was a simpler affair and largely open to the elements. The two cars could seat fourteen passengers each plus the brakeman, who was housed in a glass compartment.

The lower station of the Queens Parade Tramway can be seen on the left in this view of the pier entrance *c.* 1880.

The lift seemed ill-fated from the start when on its opening day, 8 August 1878, one of the cars broke loose and crashed into the lower station, forcing its closure for the remainder of the year. It was repaired and reopened, but the lift was plagued by frequent mishaps, including pump engine and water supply failures and landslips of the cliff. The concern was a financial disaster, and following another landslip in August 1887, it was decided to close the lift permanently. The lift was acquired by Scarborough Corporation, who removed it and laid out the cliff side as Clarence Gardens, which opened in 1890.

Above: The best surviving photograph of the short-lived Queens Parade Tramway, Scarborough, is this view taken *c.* 1880. Note the iron trestle carrying the track and the bridge over the line.

Right: The Queens Parade Tramway can be seen on the right of this photograph, which was taken on the promenade pier *c.* 1880. The pier was destroyed by a storm on 7 January 1905.

Scarborough Central Tramway 1881

Scarborough's Central Tramway runs from St Nicholas Cliff to the South Bay foreshore and is still operated by its original owning company, Central Tramway (Scarborough) Limited, which is Britain's oldest surviving tramway company. The company was registered on 20 January 1881 and the tramway was opened on 1 August 1881 at a cost of £10,358. Erected by George Wood of Hull, it has a track gauge of 4 feet 8½ inches and runs for 234 feet (71 metres) on a gradient of 1 in 2.82. The line was originally steam powered with the winch gear housed below the track 60 feet from the upper station. The lift operator had no view of the cars and relied on an indicator and other visual aids, such as string tied to the haulage ropes and chalk marks on the winch drums, to indicate the arrival of the car at the lower station.

In 1920, the tramway was converted to electric working using the 500 V DC supply of the Scarborough's electric tramway system. However, upon the closure of the tramway system, Hudswell Clarke & Company of Leeds converted the lift's supply to AC drive in 1932 using a 60 hp motor placed in the upper station. The double track was also re-laid and local coachbuilding company Plaxtons supplied new cars with seating room

The upper station of the Central Tramway, Scarborough, which doubles up as a café, photographed in June 2004.

for twenty and standing room for ten. These survived until 1975, when one of them was damaged by a fire in the adjoining Olympia building and new aluminium cars were provided by George Neville Truck Equipment of Kirkby-in-Ashfield. During the following year, the tramway suffered a further mishap when pile driving at the Olympia caused the track to subside. It was closed for a year, during which the contractors were taken to court and were ordered to pay full compensation for the repair work carried out and loss of income. In 2009, further improvements were carried out when the drive system was fully automated using a 60 hp motor.

Unusually for a cliff lift, the pay station is in the upper station, which also houses a café. The tramway is open for most of the year and has a journey time of thirty-eight seconds.

Above and right: The upper and lower stations of the Central Tramway, Scarborough, photographed in September 2015.

Scarborough St Nicholas Cliff Lift 1929–2007

Two further lifts were promoted in Scarborough during the 1920s. The St Nicholas Cliff Lift, linking St Nicholas Cliff and the Grand Hotel with the spa and aquarium, was opened by the St Nicholas Cliff Lift Company on 5 August 1929. Designed by Borough Engineer H. W. Smith and constructed by the Medway Safety Lift Company at a cost of £6,503, the double track lift ran for 103 feet (31 metres) on a gradient of 1 in 1.33 and was electrically worked by a 45 hp motor housed in the upper station. There was no lower station and passengers stepped out of the cars straight onto the pavement. The two cars could carry a total of thirty passengers and ran on a wide track gauge of 7 feet 6 inches.

The lift was purchased in 1945 by Scarborough Corporation, who provided new wooden cars in 1975. However, in February 2007, the lift was closed as £445,000 was needed to meet new health and safety requirements. The lift was mothballed, until the upper station was converted into a café, which also utilises the two cars as extra dining space.

The St Nicholas Cliff Lift, with the Grand Hotel towering behind it, photographed in the 1930s.

Above: The St Nicholas Cliff Lift, Scarborough, in operation in June 2004.

Right: The St Nicholas Cliff Lift was closed in 2007, and at the time of this photograph taken in September 2015, the upper station and two cars formed part of a café.

Scarborough North Cliff Lift 1930–96

The second of the two Scarborough lifts promoted in the 1920s was the North Cliff Lift, which ran from Peasholm Gap to Alexandra Gardens on a gradient of 1 in 2 for 166 feet (50 metres). Opened in August 1930, the lift was built as part of a large corporation development of North Bay, which included and extension of the promenade and a miniature railway. The Medway Safety Lift Company provided the lift at a cost of £8,283 and it was electrically worked by a 47 hp motor. The two cars ran on a track gauge of 6 feet 6 inches.

The lift was normally open from Whitsun to September, but was never profitable. At the end of the 1996 season, it was closed by the council, who claimed it was losing them £11,000 a year and expensive repairs were needed. Two years later, the lift was dismantled and taken to Launceston in Cornwall, where plans to re-erect it have not happened. Back in Scarborough, the site of the lift was landscaped and it's hard to visualise now that it was ever there.

Above left: A postcard of the North Cliff Lift, Scarborough, in pristine condition soon after it was opened in August 1930.

Above right: A photograph of the North Cliff Lift, showing it shortly before it was closed in 1996.

Shanklin Cliff Lift
1893–1940 and 1958

The first proposal to build a cliff lift between the seafront and cliff top at Shanklin was in 1882, and two years later, Magnus Volk, who had just built the pioneering Volk's Electric Railway at Brighton, put forward his plan for a lift. This never came to fruition, nor did another proposal in 1888. It was the financial clout of Sir George Newnes that finally came to the rescue and he presented a vertical lift to the Shanklin Lift Company in 1893. The passenger cages were raised and lowered by hydraulic power using sea water pumped by a gas engine to a 12,000-gallon tank at the cliff top. Another tank stored fresh water from springs in the cliff. The cars could hold up to a total of forty passengers and the fare was 1d down and 2d up.

During the Second World War, the lift was badly damaged when used as part of the Pipe-Lines Under The Ocean (PLUTO) operation. Fuel was stored in a large tank before being fed by gravity to a pumping station at the base of the lift, where it was pumped across the English Channel to France. The lift was demolished, but a replacement was provided by Sandown-Shanklin Council at a cost of £16,000 and opened on

A postcard of the original hydraulic lift at Shanklin c. 1905, with the lift cage at the top.

14 May 1958. This was housed in a reinforced concrete shell and the cars, holding twenty people each, were electrically powered. They took thirty-three seconds to travel the 110 feet (33 metres) up and down the cliff.

In March 2015, the lift was closed due to concerns about the condition of the metal footway at the top entrance. A temporary structure was put in place before a new bridge was provided by REIDsteel. The control system and two lift carriages were also replaced, and the lift was reopened in May 2017.

Following wartime damage, a new cliff lift was opened at Shanklin in 1958, and this postcard shows it just after it had been opened.

A photograph of the Shanklin Lift in May 2017 showing the newly constructed walkway at the top.

Shipley Glen Tramway 1895

Often thought of as a cable tramway, the Shipley Glen Tramway is actually a funicular railway. Its two cars run on a double track of 1,178 feet (359 metres) with a gauge of 1 foot 8 inches and are connected to a cable, which runs around pulley wheels at both ends of the line. Rollers keep the cable off the ground and guide it around a slight curve. The original motive power was provided by a gas engine, but this was changed to a liquid fuel engine in 1915, which in turn was replaced by electric working following a takeover of the tramway in 1928.

The tramway has seen periods of closure, and has been threatened with permanent closure several times. However, it has proved to be a doughty survivor and a reminder of when the glen was a very favoured beauty spot in the Victorian era. Amusements began to be established there from the 1870s and grew to encompass a switchback railway (moved from the 1887 Saltaire Exhibition), helter skelter, toboggan slide, aerial flight, camera obscura and Japanese gardens. Local entrepreneur and publican Sam Wilson was the driving force behind establishing Shipley Glen as a tourist attraction and he conceived the tramway as an alternative to his customers having to walk up the steep path towards the fairground. The line was opened on 18 May 1895 at a cost of £998 and had four toast-rack cars built by S. Halliday of Baildon. These had seating for twelve in each car and were coupled in pairs. Two new tramcars were introduced from 1905 to 1907, and could each seat forty-two people. The upper station, clad in corrugated iron, housed the control room, engine house and workshop, and the lower station consisted of three wooden platforms and a small pay booth.

In 1919, Sam Wilson sold the tramway to Eddie Woodhead, who in turn passed it on to Herbert and Patti Parr in 1928. They reconstructed the wooden platforms in stone and concrete and added an ornamental entrance archway at the upper station. The wooden trams were rebuilt with metal frames. Just prior to the Second World War, the Parrs sold the tramway to a consortium of Bradford businessmen, who formed a company called Glen Tramways and hired George Rushton to run and maintain it. Glen Tramway's involvement with the tramway came to an end in 1966, which proved to be a bad year for the line. In May of that year the trams overran the platform after the motorman failed to shut the power off and apply the brake. As a result, a dead man's handle was installed at the upper station, which had to be fully depressed to keep the motor supplied with power, and buffer stops were put in place at the lower station. Sadly, thieves soon stripped the tramway of its copper and non-ferrous metals, rendering it unusable. The tramway was closed, and remained so until 1969, when it was reopened by Glen Enterprises. They operated it until Easter 1981, when once again the line's future was thrown into doubt. Fortunately, the Bradford Trolleybus Association took on the tramway and, with the assistance of a grant of £10,000 towards restoration costs, they were able to reopen it on Whit Monday 1982.

The tramway is now run by volunteers on behalf of the Shipley Glen Tramway Preservation Company, a charitable trust, and is open weekends between April and December and Sundays between January and March. The two tramcars run at a maximum speed of 7.5 mph through a wooded glen over gradients varying from 1.7 to 1.20. Recent improvements included a rheostat control for speed and new braking system controls. Sam Wilson would be proud to see that his little tramway is still running 120 years on.

A tramcar full of passengers waits to descend from the top station on the Shipley Glen Tramway during the inter-war years.

GLEN RAILWAY, SHIPLEY GLEN   RELIABLE SERIES 829 / 18

A postcard sent in 1909 showing a pair of toast-rack tramcars arriving at the lower station on the Shipley Glen Tramway.

Above: The wooden glen through which the Shipley Glen Tramway runs is in full bloom on this postcard of the line posted in 2007.

Right: A photograph showing the improvements to the tramcars since 2007, which include providing cover from the elements and painting one of the cars red and the other blue.

Southend-on-Sea Reno Electric Stairway 1901–12

Developed by the American Jessie Reno in 1892, his electric stairway at the popular resort of Southend was promoted by the Reno Inclined Elevator Construction Syndicate and was built by G. Anton & Son of London. It ran for 160 feet (48 metres) on a gradient of 1 in 3 from Western Esplanade to Clifftown Parade and was covered with a wooden shelter. A 38 hp electric motor powered the wooden slatted conveyor via a chain. Opened in the summer of 1901, the stairway transported 34,853 passengers at a penny a ride in its first week, providing a net profit of £87 4s 6d. Nevertheless, the stairway was noisy and suffered frequent breakdowns, leading to its removal in 1912 (when Reno's British company was dissolved) and its replacement by a cliff lift.

In addition to the Southend and Clacton stairways, others were proposed for Ramsgate and Tynemouth. It is possible that some of the Southend Stairway was re-used in the Cunningham Camp Escalator at Douglas in 1923.

Above left: The Southend Reno Electric Stairway can be seen on the left of this postcard from *c*. 1905, running parallel to the set of steps leading up to Clifftown Parade.

Above right: A queue of customers wait to pay their penny fare to ride on the Southend Reno Electric Stairway *c*. 1910.

Southend-on-Sea Cliff Lift 1912

Following the removal of the Reno Electric Stairway, Waygood were commissioned to construct Southend's cliff lift on the same site, which was opened on August Bank Holiday 1912. A single electrically operated car runs for 130 feet (40 metres) on a gradient of 1 in 2.28 and is counterbalanced by a weight positioned beneath the running rail of 4 foot 6 inch gauge.

The lift has been rebuilt several times by its owner, Southend Borough Council, with Waygood Otis (later Otis Elevators) carrying out the work in 1930 and 1959. The 1959 work included the complete rebuilding of the upper and lower stations with cedar wood cladding, new driving and braking controls and a new passenger car. This survived until 1990, when a new car carrying eighteen passengers was installed, but in 2003 the lift was closed due to a malfunction. Full restoration, at a cost of £3 million, was not completed until 2010. The capacity of the car was reduced to twelve persons, plus the attendant, who can operate the car if needed, although the decrease in speed and final halt at the stations normally operates manually.

Passenger numbers using the lift have been affected by the opening of the free Pier Hill escalator lift. In 2017, the cliff lift was not running due to technical difficulties.

The Southend Cliff Lift was opened on the site of the Reno Electric Stairway in 1912 and can be seen here during the 1920s. The lift only had one single car.

THE LIFT, SOUTHEND-ON-SEA

THE CLIFF LIFT, SOUTHEND-ON-SEA.

Left: A postcard showing the Southend Cliff Lift, which runs for 130 feet, in 1933.

Below: The Southend Cliff Lift, photographed in September 2017.

Southend Pier Hill Lifts 2004

This observation tower and double elevator lift was installed by Southend Borough Council in 2004 to connect the high street and upper part of the town to the seafront and pier below. Designed by Stanley Bragg Architects, the tower and lifts were built as part of the £5.8 million Pier Hill redevelopment project, with the cost being met by grants from the Office of the Deputy Prime Minister and European Union Objective 2 Scheme. The lifts are open all year round.

The Southend Pier Hill Lifts and pier, photographed from Pier Hill in May 2007.

A view from the pier of the Southend Pier Hill Lifts in September 2017.

Swansea Constitution Hill Incline Tramway 1898–1902

This short-lived incline tramway was opened in suburban Swansea on 27 August 1898 and ran on a single track for 924 feet (281 metres) up the centre of Constitution Hill from St George Street to Terrace Road, with a stopping place and passing loop at Montpelier Road. The track gauge was 3 feet 6 inches and the two cars, built by Brush Engineering of Loughborough, had angled seating for eighteen passengers each and carried a driver/conductor.

The tramway was the brainchild of local businessman William Bondfield Westlake and in 1896 the Swansea Constitution Hill Tramway Order was passed. The Swansea Constitution Hill Incline Tramway Company, registered on 26 April 1897 with a capital of £11,500, engaged George Croydon Marks as consulting engineer and George Webb of Westminster as contractor, overseen by resident engineer Charles Tamlin Ruthen of Swansea. It was initially hoped to operate the line on the water-balance system, but it was decided to install a winding house at the top of the hill and drive the cable by means of a gas engine. There was no depot and the cars were to stand at the termini when not running.

In April 1898, the finished tramway was inspected by the Board of Trade representative, but he refused to give permission for it to run, stating that 'it became evident that when the lower car was fully loaded and the upper car empty, the mechanism was incapable of moving the cars'. The braking system was also found to be insufficient to stop the car in the event of a cable breaking and the crossing place switches could be tampered with. Alterations were carried out and a further inspection passed the line fit for public traffic, but on the opening day a fault developed in the winding house, which put it out of action for nearly a week.

The novelty of the tramway brought crowds of visitors to experience a ride on it, but it proved to be a nine-day wonder and the regular custom from locals never materialised. The income proved to be insufficient to cover costs and the tramway was closed in 1902 and was eventually removed, although the winding house survived until 1977.

A rare postcard view of the Swansea Constitution Hill Tramway, which closed in 1902.

Torquay Babbacombe Cliff Railway 1926

Proposals for a cliff railway to connect Cliffside Road in Babbacombe to the popular Oddicombe Beach were first proposed in 1890 when George Newnes offered to put up the finance to build it. Local objections put paid to the idea, and other subsequent schemes, until 1923, when approval was finally given. Following consultations with George Croydon Marks, work commenced in December 1924 with Torquay Corporation overseeing the project with the National Electric Construction Company and Waygood Otis. On completion, the lift was leased to the Torquay Tramways Company and was opened on 1 April 1926, having cost £15,648 to build. The Mayor of Torquay, Alderman John Taylor, made the first trip.

The electrically operated tramway runs for 720 feet (220 metres) on a gradient of 1 in 2.88 and track gauge of 5 feet 8 inches. Passenger numbers were good from the start and in 1935 they totalled 192,000. In April of that year, Torquay Corporation acquired the tramway for £18,000, and following wartime closure, spent a further £10,000 on reopening it on 29 June 1951. New cars were then provided, and further work was carried out in 1993 at a cost of £60,000. Between 2005 and 2007, the track, cars and control panel were replaced. Then, on 1 August 2009, the tramway passed to the Babbacombe Cliff Railway Company, a community interest company, who opened a visitor centre on Oddicombe beach in 2012.

The tramway carries around 100,000 passengers annually and is open all year round.

The lower station of the Babbacombe Cliff Railway and Oddicombe Beach, seen on a postcard from the 1950s.

Looking down the Babbacombe Cliff Railway towards the sea in the 1990s.

The upper station of the Babbacombe Cliff Railway in 2016.

Whitby West Cliff Lift 1931

The attractive Yorkshire coastal town of Whitby has an elevator lift linking the West Cliff with the beach below, which was opened in 1931. However, it could have had an incline tramway in the style of those at nearby Scarborough, but the proposal in 1925 was rejected for 'not keeping in appearance with the town'. In 1928 a new scheme was proposed, for a pair of elevator lifts placed in a 121-foot-long (37-metre) shaft through the West Cliff. Opened three years later, the lifts were approached from the lower promenade by a 221-foot tunnel dug by miners. The upper station, built on eight brick pillars, housed the electric motors.

In 2003, one of the lift cars was taken out of service and has since been removed. The remaining lift is usually open from May to September and can carry fifteen passengers. In 2013, the lift's bolts (more than 3,000 of them) needed replacing at a cost of £85,000.

Above left: The upper station of the Whitby West Cliff Lift, photographed upon opening in 1931.

Above right: The tunnel leading to the lower entrance of the Whitby West Cliff Lift in 2015.

Sources and Bibliography

Many written and online sources have been consulted for this book, including the individual websites for each cliff lift. Regarding written sources, *Cliff Railways of the British Isles* by Keith Turner (Oakwood Press, 2002) is the best of the previously published cliff lift books and, with 176 pages, goes into greater historical and technical depth than the space for this book can allow. Books on the Folkestone and Lynton and Lynmouth cliff lifts, Matlock Cable Tramway, Llandudno Great Orme Tramway, Shipley Glen Tramway and the Isle of Man lifts and tramways have also been consulted.

About the Author

Martin Easdown is a well-respected seaside historian who is the leading authority on the history of the seaside pier in the United Kingdom. His first book on piers appeared in 1996 and he has since had published works on the piers of Kent, Sussex, Hampshire and the Isle of Wight, Wales, Lancashire and Yorkshire. There have also been books on the piers at Folkestone, Southend-on-Sea, Scarborough, Ramsgate and Woody Bay. In addition, Martin has written about amusement parks in Britain and Warwick's Revolving Towers, and on the local history of Folkestone, Sandgate, Hythe, Pegwell Bay and Hampton-on-Sea. As a proud 'Man of Kent', he has had books published on the First World War air raids on the county, and its lost country houses.

The surviving car of the 1890 Folkestone Leas Lift, photographed in August 2017 on display at the Elham Valley Line Trust Countryside Centre and Railway Museum.